APPLYING THE SCIENTIFIC METHOD TO THE UNDERSTANDING OF EXISTENCE

John Burgener

PublishAmerica
Baltimore

First printing

Hardcover 978-1-4626-1351-9
Softcover 978-1-4626-1350-2
PUBLISHED BY PUBLISHAMERICA, LLLP
www.publishamerica.com
Baltimore

Printed in the United States of America

FOREWORD

In my university days I had a good friend who was a philosophy student. We often had discussions on science and philosophy. As a result of these discussions I acquired a life long interest in philosophy. During my business days I had little time to pursue it. When I retired my time was spent with my wife in travel and other occupations of seniors. However as age slowed me down physically, I had time to develop some of my thoughts in science and philosophy. The meaning of existence often puzzled me. So as a scientist I asked: "What is existence?"

ACKNOWLEDGEMENTS

I often had great discussions with my three sons on many matters relating to science and philosophy. As I developed my thoughts on existence Paul, Peter and John encouraged me and offered valuable suggestions. My granddaughter, Samantha, and her husband, Scott, offered grammatical improvements and proofread the final edition. I thank them all.

APPLYING THE SCIENTIFIC METHOD TO THE UNDERSTANDING OF EXISTENCE

CHAPTER 1

OBSERVATION OF EXISTENCE

Socrates and Aristotle and many since have struggled with existence. They have attempted to describe existence in terms of logic. I suggest it be treated in the same manner as the physical sciences. The physical sciences have made tremendous progress in understanding the physical world in the last two hundred years. This was accomplished by observing the phenomena around us in the world and extending it to nature and the universe in general. From the observations, theories have been developed that give us an understanding of the observed phenomena, not only on Earth but in the universe in general.

The theories do not provide a logical proof, but they have a high possibility of being correct. It is to be noticed that concrete logical proof is not possible in many cases. For example there is no possibility of logically proving the sun exists. We can see its light and feel its warmth but there is no way we can prove logically it is the sun. However by general observation, we conclude the sun exists and is responsible for light and warmth. Although not a proof it is a good working theory. In fact most would say that this is a self-evident fact. Let us do the same with existence.

What can we observe about existence? I exist and the reader must exist, because if he or she did not exist he or she would not be reading this. Let us observe the available facts of existence and then draw our conclusions. The first thing we observe is that existence can take many forms. From the simplest inanimate object such as a grain of sand to the most complicated existence that we can observe, the human brain. Although we cannot touch, see, or hear a thought, the thought does have existence. We can observe the effects of non-material existence, such as a thought or an emotion, in the same way as we can observe the effects of an existence such as light or radio waves. These and many other forms of existence do exist. The universe is full of existences, more than we can imagine. There are existences we have to observe indirectly, and can assume that there are existences unknown to us. These existences exist whether I exist or not.

One can ask if an iceberg in the Arctic rolls over with a great splash, does it make a noise? The sound waves exist whether anyone hears them or not. Some argue that if no one hears it, there is no sound. That makes existence personal. But sound is not personal per say. It is only personal to the person who hears it. The disturbances in the air are there, independent of the surroundings.

The most intriguing existence we can observe is life. Why life came into being we really don't know. However, once it became part of existence in the world it blossomed.

There appears to be different levels of existence. A rock is a piece of matter but cannot itself do anything different than being a rock. With external intervention it can change, such as chemically or due to heat. A tree has existence, but its level of existence is higher than a rock. It changes, it grows, it seeks light all within itself. However, It cannot choose an alternate way. It does not have free will. The tree follows its nature. It makes use of externals to do these things, but it does them itself. This ability to do things itself, we define as a living existence. Therefore a tree has a living existence, which is a greater level of existence than a rock. A dog has living existence, and again it has a higher level of existence than a tree. It is debatable whether a dog makes intelligent decisions, or acts in unusual circumstances only according to his nature (instincts). A human has an even higher level of living existence and can make intelligent decisions based on information presented.

The rock has an inanimate existence. It consist of matter only and posses no immaterial characteristics. The tree, the dog, the human are living existences. Living existence seems to embody something other than matter. When a living existence ceases to function (dies), the matter is still there, but something has gone, the living element. Unlike any other system it immediately starts to decay. Nor can it be repaired and brought back to life. A car breaks down it can be repaired and brought back to perform its purpose. However, parts of a living existence can be kept alive for a limited time and added to another similar living existence. It seems that something other than matter is involved. Lets call the living element the life force.

All living existences that we can observe need matter to exist. A thought however has existence, but is not matter. At first glance it appears to depend on a high level of living existence for a thought to exist. There are routine thoughts such as 'I need a drink' and there are creative thoughts such as a poem or a letter. Animals have routine thoughts such as 'I need a drink'. They do make nests and other simple dwelling places and use primitive tools. Their creative ability is very limited. However, they do have thoughts and are creative. It would appear that all living existence has the potential of being creative at the level of their living existence and capability.

What is a thought? We cannot touch, see, or feel a thought, but it does have existence. It would appear that a thought can only exist in a being with a brain capable of thinking. A rock cannot have a thought. A thought is an image and can originate in a human and possibly in other living existences. The thought image is dependent for its existence on the thinker. A poem first exists as an image. Unless written or spoken so that others see or hear it, its existence is entirely dependent on the thinker. The thinker can maintain the existence of a thought or deny it existence. For a thought to exist outside the thinker in a material world it must be shared. There are also dreams, that have existence, and even though they occur in humans and possibly in animals, dreams have no matter. There are also visions and inspirations and impressions, which occur in living matter. These may be caused by some external cause but often relate to some event or future event. These exist and again are not matter but seem to require living existence to exist.

Mental telepathy is experienced by many people. Dogs and possibly other animals also appear to experience it. It is also quite common for a thought or its result to appear more or less simultaneous in many places. This may happen because the conditions are right for it to happen. However mental telepathy does not seem to require any special conditions. In so doing a thought's existence is being shared. It can be argued that the brain sends out radiation that other brains can detect. In this way the thought, existing in the original brain is like a radio transferring the information to other brains that are tuned in. This form of radiation has never been detected, however that does not mean it does not exist. This transferring of thought could indicate that a thought although nonmaterial can exist independent of matter. As such it could be shared, even if the originator does not chose to share it. If that is the case it would indicate that we are part of a non-material world existing around us. A world that is hard to observe since our observation tools are sight, touch, sound, all material senses. There are feelings of emotion, love, happiness, sorrow and many other feelings. These feelings are immaterial and can be sensed by others. These just like atoms and molecules can only be observed indirectly. But never the less they do exist.

The most powerful aspect of living matter is claimed to be the urge to reproduce. Instead as we will demonstrate that the very essence of existence is to share. All living existence shares its existence by reproducing its existence, thus producing offspring. Trees produce hundreds of nuts or seeds. Plants produce seeds in abundance. In general living forms share their existence by producing offspring. Life in its simplest forms, such as single cells, shares it self by dividing. In other words, all life strives to share its existence. In so doing it is sharing its existence in such away that it does not diminish it's own existence. All living existence starts a single cell produced by a living existence. The cell by dividing shares itself, and is generating new existence. It appears that all living existences must come from primary ancestors.

For example we humans share our existence with our children without diminishing our existence. Without you, your children would not exist. We also observe that the offspring have an independent existence. That is both the sharing and the shared existence continue to exist whether or not the other existence continues to exist. The children do not necessarily do what the parents do, but are more or less copies of the parents for all shared living existence. This is due to the male and female DNA molecules, complex chemical compounds, which are shared. The male and female DNA's both act as a blueprint for the shared existence.

Another aspect of life is concern for one's offspring (one's shared existence). Even one's thoughts or art or other sharing of existence we attempt to protect. As we move up the scale of living existence the concern for one's offspring grows. Trees and plants protect their seeds with hard shells. Fish lay their eggs in as safe a place available, some going up rivers far from their habitat. Reptiles are more concerned and bury their eggs. Birds make nests and tend the eggs and chicks. Mammals care for their offspring till adulthood. Humans protect and love their shared existence for life and even after death. It appears that life not only wants to share its existence but it wants to protect its shared existence. To protect and wish good for anything is to love (very definition of love). So it appears that all living beings, in so far as they are capable, love their shared existence. We also note that the intensity of love for their shared existence increases with the level of living existence. The love or hate we have for others exists. These emotions can often be sensed by others without any physical contact. This may be due to physical changes in the lover or hater's appearance or actions. There seem to be cases where the physical appearance or actions are not the source of the sensation in others of the love or hate. The love we have of our selves exists.

The world is full of beauty. Beauty is well ordered as distinct from disorder and is described as an experience of pleasure, meaning, or satisfaction. Sunsets and sunrise display a canvas few if any artist can duplicate. The rocks, the mountains, the landscape, the flora, the trees, the animal life, humans are all pleasant to see. The universe as we see it, the sun, the moon, the stars is unimaginably beautiful. There is little if anything in nature that is not beautiful. One may argue that beauty is in the eye of the beholder. Actually the statement should say the interpretation of beauty is in the eye of the beholder. The beauty exists whether the beholder appreciates it or not. Since all existence as we know it is beautiful, beauty must be a characteristic of existence.

Another aspect of existence is the tendency toward complexity. Complexity is characterized as something of many parts in an intricate arrangement or interconnection. If one considers that life on earth started as a simple one cell being and evolved to the complexity of the human brain, then obviously the striving towards complexity is strong. It is not confined to life: even rocks strive for complexity. Minerals that are formed in molten lava, end up as crystals. It is also noticed that intelligence, the appreciation of beauty and the level of the ability to love increases with complexity in living existence.

All living existence demonstrates intelligence in keeping with its level of living existence. Trees seek light and nutrients. All animal life shows intelligence. Humans are very complex beings with a high level of intelligence. Humans are inclined to use their intelligence to share their existence. A beautiful painting, a complex machine, a great building, these exist as a result of someone expressing their intelligence and sharing their existence. Without the author the project would not exist. Humans are creators.

Although the ability of animals to create is very limited, they do create at the level of their living existence. Birds build nests, beavers build dams and huts. Humans create at their level of living existence. Human creations start as a thought. In fact everything done by humans in this world starts as a thought. Anything human-made required that a human preexisted the product and was accomplished by people expressing their intelligence and sharing their existence, without diminishing their existence.

Another interesting point is new living existence is formed only when the male and female DNA's are partnered and under the right conditions. For example a fertilized hen's egg is inanimate matter and remains so unless it is kept at the right temperature for a definite time. Thus it appears to be a chemical process. However man has not been able to duplicate this kind of process. The chicken that laid the egg sits on it for a period of time and it then becomes a living existence. The egg with all the right ingredients, the right temperature, for the right duration produces life. In fact similar conditions seem to be necessary for all life to begin. The chicken sitting on its shared existence is actually showing love for its shared existence, Love is frequently associated with warmth. In fact love seems to play an important role in shared existence. A rock cannot love and does not share its existence.

Cloning can be done, but it must come from a living existence and is not a matter of assembling the right chemicals. It is to be noticed that the human sperm appears to be living. It is really a mechanical device. The tail, which appears to be wiggling, is actually spinning from chemical reactions, not moving by choice. It is interesting to note that for all forms of life above the plant level, the female produces a limited number of ova. The ovum is a cell containing the ingredients of life enclosed in a membrane. The male produces sperm in large quantities, and fertilization is a hit or miss activity. The female is the source of life by producing an ovum that when fertilized divides. Even if conception is a chemical process, it is an amazing process certainly designed by a brain beyond human ability and imagination.

Our observation of existence, living or non-living, is that all existence in the world we can observe was shared. The material that makes up the world: rocks, water, air, we cannot directly see their origin any more than we can see the origin of the universe. However, all material objects that are associated with humans, are the product of intelligence and shared existence. All living existences, within their abilities, use the living and non-living existences around them for their purposes. Humans are capable of using almost anything in the world as though the world exists for them or they exist for the world. The later is not so since the world does not control Humans, as some environmentalists complain.

We are surrounded with material objects both living and non-living showing intelligence way beyond us. The complexities of the universe are beyond our imagination. Although much effort has been put into understanding the physical world and universe, we still know very little. Most plausible theories of the universe accept a beginning.

It must have always existed or had a beginning. Those are the only options.

There are those who claim the universe and everything in it happened by chance. If we consider the odds that one DNA molecule is the same as another is in the millions and there are billions of DNAs in the world. The odds of only the DNAs in the world happened by chance would be one in millions of billions, unimaginable odds. To accept chance as the cause of the entire universe not just the DNAs would require a leap of faith beyond imagination. Others say it always was. If it was always existing it would have to be timeless and to be timeless it could not be changing, since time is the measure of change. Alternatively to have always existed, it would have to expand and contract. However the observed facts do not support either theory. In fact, the evidence indicates it has been changing and expanding for billions of years, It is not timeless, it has shown no evidence of contracting and the odds are unimaginable that it happened by chance. The only reasonable explanation is the universe had a beginning. The Big Bang theory of the beginning of the universe still remains the best description of how the universe came to be.

In the above we have observed that existence can be material and immaterial. That all living existence as we know it is shared. All living existence loves its shared existence and its shared existence has an independent existence. As complexity increases the level of living existence, intelligence, appreciation of beauty and the ability to love increases. All material and immaterial creations by man occur because of someone expressing their intelligence and sharing their existence. The universe did not happen by chance or always existed. The universe and everything in it had a beginning and as such came into existence.

From our observations, everything in the world except raw mater came and still comes into existence as a result of sharing a previous existence. All creations of humans are the result of using intelligence and sharing one's existence. As done in physical science, from earthly observations the factors are then applied universally to the universe. For example, gravity we know on earth and apply it to the universe. So applying our observations of living existence as we know it on earth one would expect the same process would apply to all existence. Since there was a beginning to the universe, as we commented above and all known existence was shared there must have been something to share otherwise the whole process could not have started.

Thus a reasonable theory is that the universe and everything in it is sharing a preexisting living existence that shared its existence without diminishing its own existence. The shared existence has an independent existence and is loved by the sharing existence. Let us call the sharing existence the prime existence. From the above we conclude that existence as we know it is a sharing of the prime existence, Thus all existence is a sharing of the prime existence.

In our observations animals do have a limited intelligence and ability to create. Humans have a greater intelligence and capacity to create. In all cases the creations are not as great as or greater than the creator. In fact their creations (poems buildings etc) are at an immeasurably lower level of existence than themselves.

It is to be expected that the prime existence to be the source and creator of all shared existence, must have a level of living existence, intelligence, beauty, and complexity. Immeasurably greater than human.

The complexity, beauty and immensity of the universe is awe inspiring. The mystery of conception, the complexity of the infinitesimal, the beauty of flowers, birds, animals, even the location of the earth in the universe shows intelligence beyond us. The list of shared intelligence is endless and gives a good confirmation of the above stated theory.

I do not claim this as a proof but a good theory that has a very high probability of describing the actual origin of existence as we know it.

"I AM WHO AM" – I AM EXISTENCE, God's description of Himself

CHAPTER 2

SOME CHARACTERISTICS OF SHARED EXISTENCE

When we consider that all existence is sharing the existence of the prime existence, all existence must be in a limited way, also sharing the characteristics or attributes of the sharing existence. Let us consider some of the attributes of life. From observation of shared living existence we can recognize characteristics of the prime existence.

AWAREMESS

Certainly one characteristic is awareness. All life has some awareness. In fact the measure of living existence can be determined by awareness. The higher existences have five senses, which provide awareness of material things. Plant life also has awareness. Plants are aware of sunlight and moisture. Some experiments have shown that plants can recognize good and bad treatment. Plants have no nervous system nor do they have a brain, but they have awareness.

Animals seem to have awareness not possessed by humans. Dogs are known to become aware that their masters are in trouble with out any physical contact. Wild animals often show concern before a storm. Humans also have awareness without physical contact. Mental telepathy is well known. Twins often have awareness of pain, sickness, joy felt by the other twin. One can walk into a room and feel the tension existing in the person there. There are many well-documented occurrences, which are not physical and could not be observed by the physical senses. Awareness although aided by our physical senses does not appear to always require physical contact

The higher animals, including humans, have senses connected through the nervous system to the brain. These senses greatly enhance the ability to be aware of the world. Since the senses are connected through the nervous system, the brain is in touch with the entire body. When you become aware of a taste, an odour, or a pain, your brain has received a signal through the nervous system. However the taste, the odour, or the pain is not felt in your brain but in the affected part. The brain through a sense got a signal but did not actually feel the signal but provided information to your awareness, which made you aware. An unconscious person feels no pain, although the sore part is still sore. It is not felt because the brain is shut down and providing no information to the awareness. It appears that without a brain we have no awareness. However as mentioned above trees have no brain. They also have no physical senses, but still have awareness. The brain and physical senses which do provide a greater level of awareness in the higher living existences, seem to override the natural awareness that all living existences have. Since the sensations felt are definitely felt in the affected part not in your head, the brain must be providing information to the awareness (the mind or life force).

Consider a security system. The various sensors provide data to central point which shows on a screen the affected area. The sensors and control box tells nothing unless there is a screen or panel where the information is put into a format that can be viewed. Similarly neurons moving about in the brain provide the signals but need a control panel to be aware. The information is displayed or made aware by the mind. If we consider sight, the eye sees a very small image of the scene being viewed on the retina and provides this information to the brain. However we see with the same eye, a full scale image. The brain presents information to the mind and the mind forms an image. This image has existence and is immaterial, and exists in the mind not the brain. If you dissect the brain you will not find the image.

Human Level of awareness far exceeds any other living existence. The characteristics that make us human, such as free will, ethics, love, all require awareness. In fact without awareness it would be impossible to love. Humans have reached a very high level of shared existence and therefore share more of the prime existence's attributes. We can also actually be aware of the prime existence, although for most in a limited way. Some people experience an intense awareness of God.

At the human level of sharing awareness of the prime existence, we are aware of ourselves. No other existence in the world has this ability to our level. This level of awareness, to be aware of the prime existence and a self-image, makes us human. Animals have an awareness of themselves and thus a mind. They are aware of pain, love, and many other things. They do not in general recognize themselves in a mirror or photograph and as such have a limited awareness of self. It is claimed that some primates recognize then selves in a mirror. Probably recognizing their hair or face or other characteristics but this is a recognition of a characteristic of self, not a full image of self. A self-image is a full image of ones self. It allows you to try to be more than you are. Many humans do so, There is no evidence of an animal doing so. Your self-image is an actual image of what you think or believe you are. How does one make a decision? You do it by considering how it will affect you. To do so you need a knowledge of what or whom you are. A characteristic such as your hair colour or height except in limited cases would not suffice. You need a full image of who and what you are, a self- image. I believe that without a self-image it would be impossible to have a free will. It is impossible to make personal decision if you don't know yourself.

The self-image indicates a high level of awareness. Our decisions, our attitudes are all coloured by whom we are. We can exchange body parts, transplants, even lose parts and remain whom we are. In fact all our actions, free will decisions, our hates, our loves our creativeness are due to our awareness and governed by whom we think we are. The brain provides information so that the self-image can decide how the information fits our image. Our self-image, which is a function of our awareness, defines us as a person. It is an image, has real existence, and is not material. This awareness, or mind is the life force that appears to exist in all living existences. It can not be located in our material body but is somehow connected to it. As immaterial it can not be part of it. The mind holds our self-image and awareness of the prime existence. Our self-image and awareness defines us as an individual human person. Since it is immaterial it can exists independently of the physical body. Therefore it could continue to exist after the death of the body.

LOVE

Love is common to all living existence according to its level of existence, as demonstrated by love of offspring. Although considered an emotion, it is more than an emotion. Emotions in general are personal feelings, not generated by logic, that are concerned with how one feels or reacts, but not of concern for others. For example, Fear, usually, is very much concerned for ones self. If it is concern for others it really is love. Happiness is enjoyed irrespective of others. Others may make you happy but do not have to be happy themselves. You may be happy someone is happy but really that is concern for someone else, that is love. Love is a concern for the one loved. It also is love of self, since one is concerned with ones welfare. All living existences show concern for their offspring. The higher animals and humans show love for more than their offspring. The dog loves its master. Humans love not only their offspring but many others throughout life, some even after death. It does produce emotional feelings. However, as mentioned, it is more than an emotion. One chooses to love as an act of the will. Love at first sight so called requires knowledge of the loved and a decision that you want the loved one.

It is natural to love. There is a natural striving to love and be loved. However, to love requires action on one's part, a decision to wish good. Intervention by our material existence. Material interests interfere and restrict our capacity to love. This material interference can result in hate, wars, and conflict of all kinds. This is clearly shown in conflicts over justice, where a little love could easily settle the conflict. People who have developed the capacity to override the material attitudes and accept the natural inclinations of shared existence are those who do great good for the world, such as Mother Theresa and many like her.

Love is stated above a characteristic of shared living existence. The fact that all living existence is sharing the prime existence demonstrates the love of the prime existence for all creation. Considering the many gifts given to humans, including the beauty of nature and the awareness to enjoy it, the prime existence must have a great love for us. For humans thanklessness and unreturned love are painful and must be so for the prime existence.

COMPLEXITY

As stated in chapter 1 all living existence appears to strive for increase of complexity. As complexity increases the level of living existence increases. If we consider complexity in living existence we recognize that the complexity is less chaotic and more organized. It is more ordered. That is of a higher order of existence. Even the crystals that form from molten lava exist as a higher order of existence then the unorganized lava. Thus one of the characteristics of existence is the pressure to a reach a more organized existence. This is also evident in the interaction of matter in general. A mixture of hydrogen and oxygen given the right conditions will form water, which is a higher order of existence than the mixture of the two gasses.

However, if the conditions are severe enough, matter has a tendency to chaos, the crystallized rock will return to molten lava with enough heat, Matter in general will interact in either direction depending on conditions. Many carbon based materials given the right conditions will interact to produce very complex highly organized products. DNA will reproduce itself in the laboratory. It does not produce a living being, but it is a very complex reaction. Shared existence strives for more ordered and less chaotic existence.

BEAUTY

The sun is appreciated for its light and warmth but still exists whether or not one sees it's light or feels it's warmth. Similarly, Beauty, although claimed to be in the eye of the beholder, has existence independent of the beholder. Beauty gives a sense of comfort or meaning to the beholder. There are degrees of beauty and different beholders may have different responses but the organization and order that results in beauty is there. All existence that we observe is beautiful in some way. The earliest forms of life were beautiful at their level. A modern tree or flower is more beautiful than the first forms of plant life. Similarly the mammals are more beautiful than the dinosaurs. Humans are even more beautiful. From observation both visual and archeological, as the level of existence increases, beauty also increases. Beauty is a characteristic of shared existence. As such there is a striving for an increase of beauty

CHAPTER 3

EVOLUTION AND SHARED EXISTENCE

If we consider the attributes of shared living existence they include the striving to share one's existence (offspring), the striving for order, awareness, beauty, and the capacity of loving. All of these attributes lead towards a greater sharing of the prime existence. We see the striving for order in inanimate matter. Chemical reactions under normal conditions tend to a more organized state. Of course this tendency to a higher order of existence can be interfered with by unusual conditions. However all existence, even inanimate matter not only living matter, has in a limited way the attributes of the prime existence. Although we do not know from observation how life actually started, we know that all existence is sharing in the prime existence. Certainly one of the attributes of the prime existence is life. It is recognized in evolution that mutations do occur. A mutation is a change in the DNA that leads to the offspring being different than the parent. In many such changes there is no observed explanation. Because of the sharing of the prime existence's attributes, one can expect that mutations will occur when conditions allow the shared attributes to react. Especially when the tendencies to be more aware, organized, and beautiful would result in a higher order of existence. If the attribute of life were sufficiently present and conditions complex and organized enough, then a primary mutation resulting in a basic form of life could have occurred. If this were the case then life could occur in many places and many forms.

Thus if life occurred from a chemical reaction it would be a mutation. It is possible that several mutations were needed before the first living cell happened. For example viruses are not quite living since they can not reproduce. A cell is composed of the essential ingredients for life enclosed in a membrane. It would also appear that once a reproductive cell occurred awareness was also generated, since all observed living existences have awareness. Once this occurred, reproductive life started. From then to the primates and hominoids life progressed by various processes and mutations. The pressures on matter to move to a higher order of existence continued up to the Neanderthals and Cro-Magnons. The Neanderthals were very human like but not human. They were shorter and more stooped and it appears that they were less intelligent than humans. The Cro-Magnons followed and overlapped the Neanderthals. The Cro-Magnons were anatomically equivalent to humans.

It is assumed that the Neanderthals and Cro-Magnons showed an increase in intelligence over the primates. In all cases of archeological evidence of intelligence, one is looking at capability and not necessarily intelligence. All mammals are intelligent. A dog is quite intelligent but is limited in what it can do. It has no hands and no fingers. It tries to use its paws but they are not good hands. It also can not stand upright or carry anything with it's arms. These are serious physical limitations. The primates, especially the Chimpanzees, are claimed to be more intelligent. If one watches Chimps they act like most animals living in herds, however they do have hands and fingers and are partially upright, which are huge physical advantages. Wolves who survive in more difficult environments, have to hunt for food, and do so in organized packs have none of the physical advantages of Chimps. Many animals show surprising intelligence but their activities are limited by their physical limitations.

Similarly with Neanderthals, they had hands with opposed thumbs and walked upright. A dog pulls things with his mouth but a Neanderthal could pick it up and carry it. He could move stones around and hang onto a stone to chip it with another stone. They were around for more than 200,000 years but they never moved out of the Stone Age. The mutations that produced them were physical with little if any improvement in intelligence. One wonders if the intelligence level of all mammals, other than humans, is approximately the same.

The Cro-Magnons who were anatomically similar to modern humans were around for more than 50,000 years and in some areas overlapped the Neanderthals. They were taller, more upright, and possibly stronger than the Neanderthals. They had the advantage of learning from the Neanderthals. They did grow some food and were more settled than the Neanderthals, They also appeared to draw simple pictures of hunted animals, probably as communication to other hunters. This would indicate that they could identify and form an image of the hunted. A dog or for that mater any mammal can identify its master or its offspring and many other things. To do so they must be able to form an image. They have the mental capacity to make drawings without the physical capability. I do not consider such sketches as greater intelligence than other mammals. They did have the spark to do it, never the less they remained in the Stone Age for at least

35,000 years. However, about 15,000 years ago there was a sudden increase of capability or intelligence in the archeological record. Instead of just moving stones about there was evidence of structures that required more than pilling stones, the Neolithic revolution. By 8,000 years ago structured buildings and temples were being built in Asia. In the Euphrates Valley the Bronze Age had began by 3500 years ago. This all happened in less than 12,000 years whereas for 35,000 years little had happened with the Cro-Magnons.

The obvious answer is that another mutation occurred as usual in a small population and a local area, probably about 16,000 years ago. The offspring from this mutation spread out and showed the Cro-Magnons how to use their abilities intelligently. Since modern man is anatomically the same as Cro-Magnons, the mutation was not physical but mental. It was evidently a huge increase in intelligence. It is also significant that prior to the Neolithic revolution there were no temples built. The lack of temples or places to worship indicates that prior to this mutation there was no awareness of a superior existence. The distinguishing factor between the Cro-Magnons and modern humans is a great increase in awareness, the ability to be aware of themselves and something greater than themselves. To be aware of the prime existence, to know God. This level of awareness put us on an entirely different level of existence. It is interesting to note that all current primitive peoples recognize a superior existence and are therefore modern humans. It is also interesting to note that according to mitochondria DNA data all humans are descended from one woman. This of course ties in with the original theory that all living existence is shared. It also confirms that the mutation occurred in an individual man and woman, Adam and Eve.

Modern humans, because of their greatly increased awareness, live on a much higher level, enjoying free will, understanding of life, philosophy, science, a self image which defines the individual and awareness of the prime existence. They have many other capabilities such as the ability to not only share their existence but to share it creatively in great structures, literature, art, science, philosophy and organized living. Fantastic progress occurred in less than 16,000 years. The Neanderthals for more than 200,000 and Cro-Magnons for at least 50.000 years produced little progress. The mutation from Cro-Magnon to True Human was different than other mutations, It was mutation of the mind and not physical like the mutation from Primate to Neanderthals to Cro-Magnon. It produced an entirely different species.

The mutation that produced humans actually fits the Genesis account. In Genesis, Adam was formed out of the earth. Genesis does not say that God acted like an artist and molded some clay. Instead it just says that God formed Adam from the earth. I would interpret it as God using evolution to develop animals and Adam from earth's elements, which is what Evolution also says. God breathed life into Adam. A mutation, due to the attributes of shared existence, that gave him awareness that was beyond all previous worldly existences. A mutation that certainly was the breath of God. Eve was of course a shared existence and may have been a daughter of the Cro-Magnon Adam, thus taken from him. The account mentions the inter marriage of humans the "sons of God" and the Cro-Magnons, the "daughters of man". The new humans obviously were aware of God. The Garden scene certainly calls attention to the ability and obligation to know God. The tree of knowledge of good and evil, I would interpret as living for God or living for one's self. It is interesting to note that those who live for God, like Francis of Assisi and Mother Theresa, do lead a life filled with love. The Genesis account is remarkably accurate, considering it must have been written thousands of years later by and for people without our knowledge.

In Conclusion, this concept of all existence sharing God's existence gives us an understanding and the value of all existence. We see that humans are privileged with an awareness that allows an awareness of God and a self-image, which is independent of matter and is immortal. The sharing of God's Existence and an immortal soul gives us reason for our existence: To know, love, and enjoy our shared existence and the infinity of God, and to share our existence with others.

CPSIA information can be obtained at www.ICGtesting.com
Printed in the USA
244113LV00003B/155/P

9 781462 613502